Progressive Function Point Analysis
Advanced Estimation Techniques for IT Projects

Achieve greater accuracy through Progressive Estimation
Extended functionality for FPA to include Workflows
Cost Justification based on development effort.
Support for Functional Reuse and Application Component Reuse
Prepare for CMMI Assessment & KPMG Audits.
Performance Improvement Guide
Compute both cost and schedule index from FP Count
Free open source tools and Workbooks to get you started

Ruben Gerad Mathew
Email: rubengerad@gmail.com
Mobile: +918105093030

Disclaimer

The content presented herein is not part of standard IFPUG Function Point Analysis however; we hold the IFPUG FPA principles as the basis for Progressive FPA. The progressive estimation techniques use coefficient ranges compared to standard IFPUG static FP value ranges, and additionally introduces the concept of functional Reuse and Process/Workflow integration to factor in all aspects of development.

This book is dedicated my dear God who makes all things possible, by the working of his spirit, through Jesus Christ, my savior, redeemer, friend and guide, my able teacher and my source of wisdom who blessed me with all that I have and even more, a loving wife, healthy children and a great life.

Preface

Proper estimation is crucial to the success of any project. Progressive Function Point goes a step further to help to justify the cost of a project based on actual work involved and can provide an accurate cost and schedule index for the project with limited variance. This book is gradual self-paced tutorial and guide focused on beginner to advanced levels. Some key concepts are derived from UML modeling, so knowing UML is good to have or at least knowledge of basic flow-charting principles is required. You should also be familiar to project cost accounting and must have worked in at least a few projects to be comfortable working with Progressive FPA.

Guidelines

Beginners: Start with Chapter 1 and work your way till Chapter 3
FPA Analyst: Begin with Chapter 2
Performance Tweaking: Chapter 4 is only for teams and Senior Analysts with considerable experience in FPA and needs to align the workbook for Practice Audit.

An Open source Initiative

All templates and workbook required are provided free of charge and downloadable from sourceforge.com Please do not share your personal EBook copy, you are entitled for an early release copy of the current workbook. Kindly contact rubengerad@gmail.com to obtain your free copy.

About the Author

 Ruben has been in the IT industry well over 16 years is a Senior Architect with more than 8+ years hands on experience. He has worked in travel, telecom, airlines, insurance and healthcare domains and has trained business analysts, architects and project managers in implementing FPA in projects. He has managed teams both onsite and offshore and worked with Fortune 100 clients. He is married to Anna, and is blessed with a son Jacob.

Ruben works as a freelance consultant, and provides training and support both online and onsite for Architects, Project Managers and Business Analyst.

This page is intentionally left blank.

TABLE OF CONTENTS

FUNCTION POINT ANALYSIS

Function Point Analysis is a method of estimating the size of a project by considering the input and output elements that are in the project and consolidates each type of operation into data or transaction function. The size of projects used to be computed using the KLOC (Kilo Lines of Code), but could not be applied before the project was completed, as the prediction models were far from being accurate, but the concept had similarities to FPA as observed by Allan J. Albrecht, the inventor of FPA. The FPA methodology for sizing software was devised by him at IBM. The introduction of FP Counting practice helps to ascertain the size of the project by considering all the variables in the equation to deliver the function point count, and provided better estimates than those computed by analogy and user experience.

The Function Point Analysis estimation methodology validates the individual elements and the related groups to arrive at a complexity level of high, medium and low and assigns a function point count for each subset. There are five basic function types Internal Logical Files (ILF), External Interface Files (EIF) in data functions and External Inputs (EI), External Output (EO) and External Query (EQ) in transaction functions. The elementary variables in functions are denoted as Data Element Type (DET). The functional complexity is computed as the total number of user identifiable groups that exists within DETs and is termed as Record Element Type (RET) in Data Functions and all referenced file types are counted as File Type Records (FTR) in Transactions Functions. A corresponding matrix holds the reference function point values for all function types namely the ILF, EIF, EI, EO and EQ, with respect to the range of DET and RET/FTR in each function. The total sum of the high, medium and low count of all operations is the unadjusted function point count. There are 14 General System Characteristics used to ascertain the Adjusted Function point count. The GSC computed in each project is called the Value Adjustment Factor. This is used to calculate the complexity of environment, task and language and tweak the final count for the particular environment.

The Project count can be broadly classified into three types as defined in the Function Point Manual by David H. Longstreet.

Development Function Point Count: Function Points can be counted at all phases of a development project from requirements leading up to implementation. This type of count is associated with new development work and may include the prototypes, which may have been required as temporary solution, which supports the conversion effort. This type of count is called a baseline function point count.

Application Function Point Count: Application counts are calculated as the function points delivered, and exclude any conversion effort (prototypes or temporary solutions) and existing functionality that may have existed.

Enhancement Function Point Count: It is common to enhance software after it has been placed into production. This type of function point count tries to size enhancement projects and is counted as sum of all added, changed or deleted function points in the application. By tracking enhancement size and associated costs, a historical database for your organization can be built. Additionally, it is important to understand how a development project changes over time.

The Fundamental Concepts of FPA

User

Any person, device or process that communicates and interacts with the software at any time is termed as user.

Application Boundary

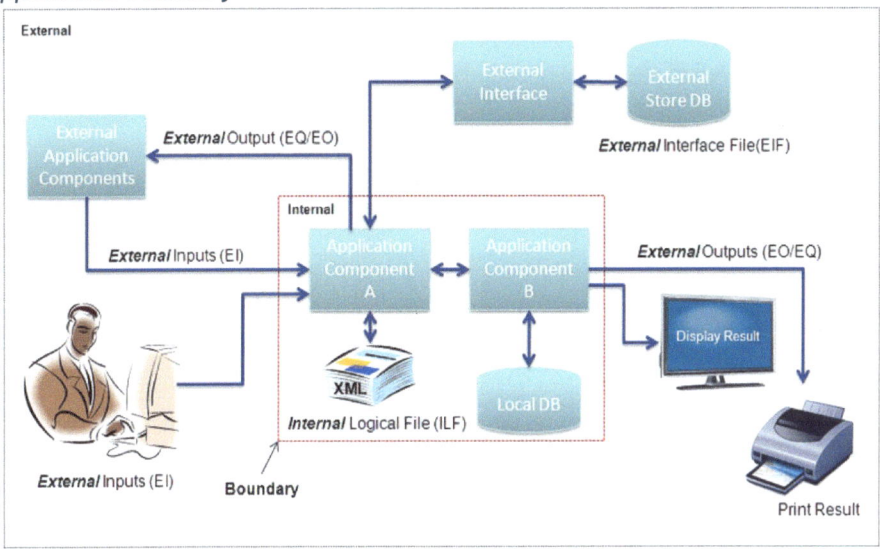

To perform function point analysis, we have to first define the boundary of the application. A boundary is used to define what is internal to the application and what component or interfaces are external to the application. This distinction is important and once the boundary is established, it remains constant throughout the lifetime of the application, and is used as reference for all future estimations. The boundary is a user perspective based on the understanding of the system and related environment.

User Recognizable

Any requirements, operation, process or realm that is understood by both the developers and business users and is commonly accepted is termed user recognizable.

User View

It is the description of a business function, operation or process which is used to calculate function points. It is the unit of work described verbally by the user and measured by variables and related processes steps.

Control Information

These are key fields that helps in determining how to process the data contained in it such as markers for historical data mining, or an indicator to show the operation is to be performed while processing the data.

To perform function point analysis the functional categories are derived into two basic types which are data functions and transaction functions.

Data Functions

Data functions relate to the actions of storing and retrieving data in both local files or databases and external to the application through remote interfaces, associated middleware or outside the boundary of the application concerned.

IFPUG: Internal Logical File (ILF)

User recognizable group of logically related data or control information maintained within the boundary of the application being measured. The primary intent of an ILF is to hold data maintained through one or more elementary processes of the application being counted.

Simple Definition

The entity model to which data is persisted (stored) by the application should be within its boundary, which may be the local database, file data source or any other type of persistence mechanism. Data is organized into groups and subgroups. The subgroups are more commonly related to tables in a database. A group is a collection of tables that has affinity to a particular entity, department, location or any other user identifiable selection criteria like Customer Records or Orders. The data is maintained in the ILFs through an elementary process, which is (CRUD) Create, Read, Update, and Delete operation from within the application boundary and is referenced by an Add, Update, Delete transaction function. Control information may be passed as an indicator of the process to be performed.

Counting Rules

The group of data or control information is logical and user defined.

The group of data is maintained through an elementary process within the application boundary being counted.

Identifiers

Validate any CRUD operation performed on a database, table, file or component including in memory caching within the boundary of the application.

IFPUG: External Interface File (EIF)

An external interface file (EIF) is a user identifiable group of logically related data or control information referenced by the application, but maintained within the boundary of another application. The primary intent of an EIF is to hold data referenced through one or more elementary processes within the boundary of the application counted. This means an EIF counted for an application must be in an ILF in another application.

Simple Definition

Data transmitted or persisted outside the application boundary, which can be accessed through the use of a remote application interface, a remote server, or data which may be maintained by another application and managed through remote interface. Each of the fields required may be individually counted and reference keys and structures may also be counted. The nature of storage and retrieval alter the form, such as into key value pairs, or serialized to a string, or exchanged as XML data. The groups and subgroups are derived from the logical structure and not data in its physical state.

Counting Rules

Count a DET for each unique user recognizable, non-repeated field maintained in or retrieved from the ILF or EIF through the execution of an elementary process.

Count only those DETs being used by the application being measured when two or more applications maintain and/or reference the same data function.

Review related attributes to determine if they are grouped and counted as a single DET or whether they are counted as multiple DETs; grouping will depend on how the external process use the attributes within the application

Count a DET for each attribute required by the user to establish a relationship with another ILF or EIF.

Identifiers

Any operation invoked through a web services interface, or remote application interface through some middleware technology or remote access through various protocols. The data is stored outside the boundary in another system typically in another application server, content management system or file server but referenced and used by the application concerned.

Transaction Functions

Transaction functions relate to read/write operations performed on the data. The transaction functions will read or write data from and to an ILF or EIF. There are three basic type of Transaction functions which comprises of External Input, External Inquiry and External Output.

IFPUG: External Input (EI)

An external input (EI) is an elementary process that processes data or control information that comes from outside the application boundary. The primary intent of an EI is to maintain one or more ILFs and/or to alter the behavior of the system. Data entering the application boundary either by user input, data feed or an external application invocation.

Simple Definition

It denotes any data input screen or operation that is used to perform a CRUD operation on the ILF (database or persistent storage). This data received has to be stored to the database as tables or to files or some persistent storage mechanism and links to the corresponding ILF. It should alter the behavior by going from input data state to an updated data state.

Counting Rules

Count one FTR for every read operation from ILF within the boundary or one per EIF file.
Count one FTR for every write operation to an ILF within the boundary of the application.
Only count the same ILF once for both read and write operations.

Identifiers

User providing data in an online application form, selecting options, choices, inputs text, uploads files or feeds data into the system. An external application sends data to be stored/processed or both. Receive data from external sources as stream feeds at intervals. Receive a device event such as alarms or triggers with relevant data to be processed.

IFPUG: External Inquiry (EQ)

An elementary process that sends data or control information outside the boundary. The primary intent of an external inquiry is to present information to the user through the retrieval of data or control information. The processing logic contains no mathematical formula or calculations, and creates no derived data. No ILF is maintained during the processing, nor is the behavior of the system altered.

Simple Definition

Denotes a simple data read operation, where information is retrieved or data is sent outside the boundary, which may simply be a display devices such as a PC monitor or information sent in response to a remote procedure call from another application outside the boundary. There may be some operations leading to the output which makes EQ and EO have both input and output parameters.

Counting Rules

Sends data or control information external to the application's boundary.

For the identified external process, one of the following must apply

Processing logic is unique from the processing logic performed by other EQs for the application.

The set of data elements identified are different from other EQs in the application.

The ILFs or EIFs referenced are different from files referenced by other EQs in the application.

Identifiers

Data that is fetched from the database displayed as a result of some user action. Data which may sent to external devices such as printers or other devices. Data which is sent to other external applications outside the application boundary.

IFPUG: External Output (EO)

An elementary process that sends data or control information sent outside the application's boundary and includes additional processing beyond that of an external inquiry. The primary intent of an external output is to present information to a user through processing logic other than or in addition to the retrieval of data or control information. The processing logic must contain at least one mathematical formula or calculation, create derived data, maintain one or more ILFs, and/or alter the behavior of the system.

Simple Definition

This is a data read operation and the result may be displayed, printed, transmitted to an external device or application outside the application boundary and is similar in all respects to EQ. The control information may be additional data that may include information on how the data may be processed. The main difference of an EO from an EQ is that it may contain some mathematical equation, calculation of sum, average, count or other manipulation of data, or may create additional derived fields such as totals, subtotals, calculation of final cost and may also update the ILF to reflect the computed sum.

Counting Rules

Sends data or control information external to the application's boundary.

For the identified external process, one of the three conditions must apply

Processing logic is unique from the processing logic performed by other EOs for the application.

The set of data elements identified are different from other EOs in the application.

The ILFs or EIFs referenced are different from files referenced by other EOs in the application.

Identifiers

Data is fetched from the data source (ILF) and some additional logic is applied to transform the data. This may result in creation of new fields which may be the outcome of some mathematical calculations and also may be saved after the information is computed back to the data source.

IFPUG: Data Element Type (DET)

A DET is a unique user recognizable, non-recursive field. A DET is information that is dynamic and not static. A dynamic field is read from a file or created from DET's contained in a FTR. Additionally, a DET can invoke transactions or can be additional information regarding transactions. If a DET is recursive then only the first occurrence of the DET is considered not every occurrence.

Simple Definition

Data element types are the granular elements or fields collected from each of the transaction and data functions. These are the attributes contained in the tables, fields in UI forms, parameters passed in application calls. It is may be data that is unique and user recognizable, for a data table we may reference the unique column headers, for form elements we count each of the form fields. Only the first unique instance is counted even if tables or records are repeated. By assessing the DETs in relation to the RETs in data function or FTRs in transaction function we derive the complexity and arrive at the unadjusted function point count. Screen elements that are static like system time or page counters are not counted as DETs. For every application, any one-action link to submit the form contents will also be counted as a DET.

Counting Rules

Count one DET for each data input field, error messages, calculated values & a single action button or link leading to submission for External Inputs.

For External Outputs count each Data Field on a Report, calculated values, error messages, and column headings that are read from an ILF.

Count one DET for EQ and EO for both input side and output sides. The input side which may be field used for selection or search, and output side is displayed fields on a screen.

Identifiers

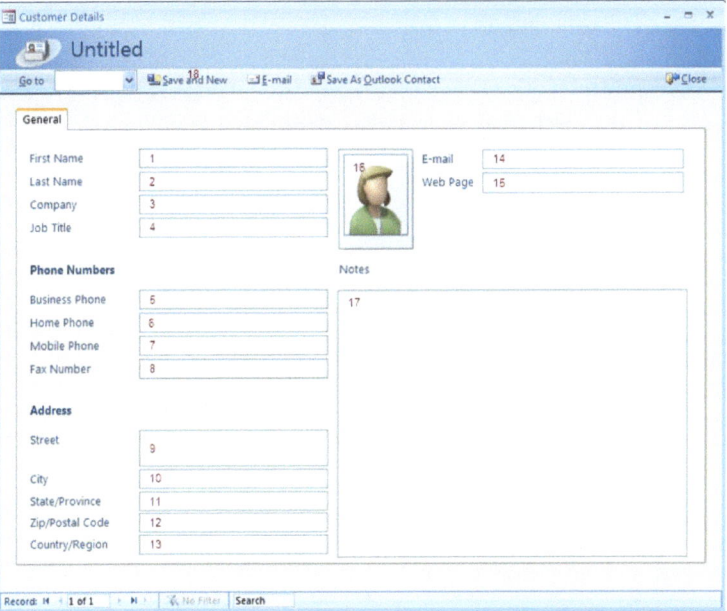

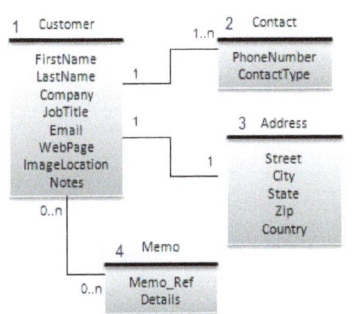

In the data entry screen shown above, we have counted 18 DET elements, of which the last element is the action button which can trigger an action. Only a single action button is computed in the equation. This transaction may be titled as "Process and Store Customer Details" and will contain 18 DETs and 1 FTR, the Customer group.

Data exists in groups and subgroups which may have sets of related data which may relate to each other when there is a direct or indirect reference field(s).

IFPUG: Record Element Type (RET)
A RET is user recognizable sub group of data elements within an ILF or an EIF. The child information is a subset of the parent information.

Simple Definition
Record Element Type (RET) is the logical sub groups of elements, which may exist in the ILF, and could be related tables in a database or relations of a parent cluster. An entity may consist of several closely related tables or composite relations, for which one RET is counted, for other relational tables where they are related by common key such as in an association relation, count one RET for each referenced table, file type or entity. The complexity measure for a given function is weighed by counting the number of RETs identified in the DETs.

Counting Rules
Count one RET for each table in a user identifiable group in ILF or EIF.
Count one RET for each sub-group in a parent group.

Identifiers

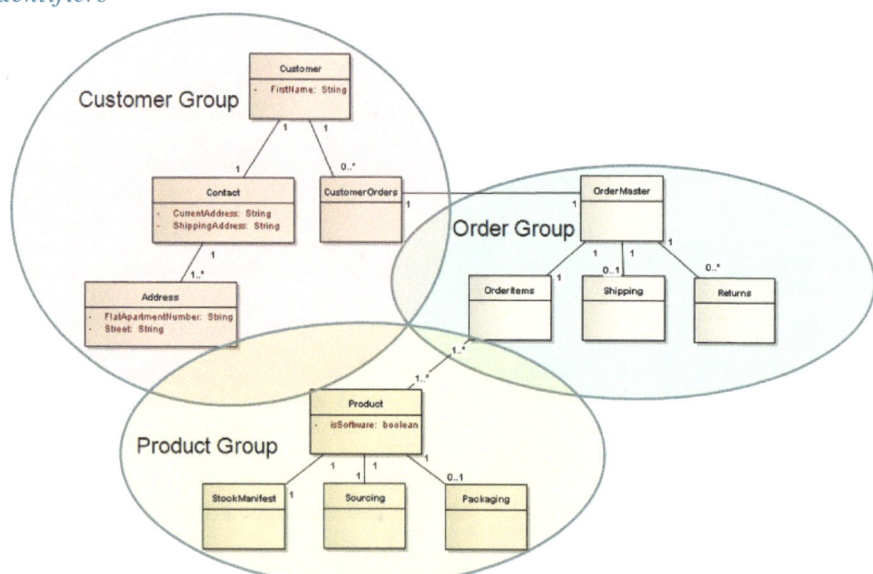

To identify RETs, we must first identify the respective user identifiable groups such as customers, orders, products. In each group there will be sub groups or related tables within the group. We count the number of subgroups in each group to list the number of RETs we are dealing with in a particular function, and count one RET for all the related tables from other groups which are referenced.

IFPUG: File Type Referenced (FTR)

A FTR is a file type referenced by a transaction. An FTR must also be an internal logical file or external interface file.

Simple Definition

File Type Referenced are logical groups either local storage or remote identified in a Transaction function. They are logical groups of data which must either be identified as an ILF or EIF.

Counting Rules

Count one FTR for each ILF or EIF in the transaction function.

Identifiers

Identify all the ILFs and EIFs in an application, and follow the process steps in a transaction function and count each unique reference to the EIF or ILF.

Reuse FP

Reuse FP is important to denote the functional reuse that may exist within a project. There are two types of Reuse implemented in general, the functional reuse is used with each data and transaction function to denote the percentage of reuse in the given function, this is not a part of the standard IFPUG practice and is used only in progressive function point estimation. There may not be sufficient time in estimating the size of reusable modules when they are large complex libraries, or may not be available in certain COTS products when they are designed as a black box, for such instances, the weighed reuse percentage is based on user experience and analogy. A 0% reuse represents 100% new software build, any percentage denoted shows the total amount of reusable components available to the application for integration.

Reuse % = Reusable FP Count/Total FP Count

The second method of reuse is calculated with Adaption Adjustment Factor (AAF) was presented in an IFPUG conference to calculate reuse with respect to integration of an entire application or component as a whole for any project. It takes into consideration three factors Design Modification (DM), Code Modification (CM), Integration & Testing (I&T).

AAF = .4DM +.3CM +.3I &T

Steps in Calculation of Reuse count
- Calculate FP count of Reusable component. (100FP)
- Determine the percentage of Development Effort. (50%)

- Determine the percentage of Coding Effort. (25%)
- Determine the Testing & Integration Effort. (75%)
- Calculate Adaption Adjustment Factor

AAF = (0.4 x0.25)+(0.3x0.50)+(0.3x0.50)

AAF =   .1   +   0.15   +   .15

AAF =   .4

Calculate Reuse

Reuse = FP Count of Reusable component x AAF

Reuse = 100 x .4

Reuse = 40FP

Counting Function Points

Unadjusted Function Point Count

Unadjusted function points refer to total counted FPs in a project. The FP estimation begins after all data and transaction functions are keyed in. The function point value for each Data and Transaction function is identified for operations Added, Changed and Deleted functions that are in a given User Story or Use Case. The function point count is based on the number of DETs and RETs/FTRs in the matching row and column of the Function Point Reference Matrix. The FP Reference Card provide value ranges for Low, Medium and High complexity transaction FP count based on the DETs and RETs present in a Data function or the DETs and FTRs in a transaction function.

ILF - Value Ranges			
	DET		
RET	1-19	20-50	50+
<2	1	7	7	10
2,5	5	7	10	15
>5	6	10	15	15
Range		20	51	51

EIF - Value Ranges			
	DET		
RET	1-19	20-50	50+
<2	1	5	5	7
2,5	5	5	7	10
>5	6	7	10	10
Range		20	51	51

The FP value table is the IFPUG reference card and used for calculating the FP Count of ILFs and EIFs in Data Transactions. There are 3 ranges for DETs, and RETs. Based on the each of the Low, Medium and High categories, a specific FP value is provided for data and transaction functions. This forms the basis of FP count determination. The total count derived from all the data and transaction functions listed in the project as the Unadjusted Function Point value.

Adjusted Function Point Count

The complexity factor can influence the final project count by +35%/-35%. The final count is termed as the adjusted function point count.

Value Adjustment Factors are used to derive the final adjusted function point count of an application. It uses 14 General System Characteristics (GSC) to identify the total complexity factor for the application and are classified into the following categories:

- Data communications
- Distributed data processing
- Performance
- Heavily used configuration
- Transaction rate
- On-Line data entry
- End-user efficiency

- On-Line update
- Complex processing
- Reusability
- Installation ease
- Operational ease
- Multiple sites
- Facilitate change

VALUE ADJUSTMENT FACTOR (VAF)

Value	Characteristic
0	Not Present, No influence
1	Incidental influence
2	Moderate influence
3	Average influence
4	Significant influence
5	Strong influence throughout

#	Degree of Influence	Value (0-5)	Comments
			The VAF can be adjusted as per project env, by setting the VAF to 1.00 ensures no VAF is applied on the final FP count.
1	Data Communications	0	
2	Distributed Data Processing	5	
3	Performance	0	
4	Heavily used configuration	5	
5	Transaction rate	0	
6	Online data entry	5	
7	End-user efficiency	0	
8	Online update	5	
9	Complex processing	0	
10	Reusability	5	
11	Installation ease	0	
12	Operational ease	5	
13	Multiple sites	0	
14	Facilitate change	5	
	Total	35	
	Value Adjustment Factor (VAF)	1.00	

For each of the GSC a rating of 0 to 5 can be provided which represents

0=Not present, or no influence

1= Incidental influence

2=Moderate influence

3=Average influence

4= Significant influence

5=Strong influence throughout

Once all the details of the project are provided, and the VAF is calculated, the Unadjusted FP is multiplied with the VAF factor to arrive at the Adjusted Function Point Value. The VAF can influence the project by increasing or decreasing the unadjusted FP count by 35%. This ensures that the complexity of the environment is factored into the project.

PROGRESSIVE FUNCTION POINT ANALYSIS

Progressive function points was derived for greater accuracy in the estimation of Function Points using the same FPA principles but attempts to improve on the actual Function point count based on elementary inputs, outputs and integrated process flows within the respective functions. Progressive function points are better suited for application that may have the following characteristics

- Consistent measures above IFPUG FP Reference range.
- Complex operations other than standard CRUD functionality.
- Integration of Reuse within functions.
- Greater Accuracy of Cost & Schedule Index.
- Greater Clarity and Accountability into costing of applications.
- Passing external reviews such as CPMG Audits.
- Integration of new languages and programming tools.

The elementary process steps of a function are derived from a sequence diagram or activity diagram. The sequence diagram may contain the depiction of both data and transaction functions.

Definition

Progressive Function Point Analysis is a method of estimating the size of a project by considering the input and output elements and *the elementary process steps that are in the project*. The estimation methodology validates the elementary input, output elements and *elementary processes* with respect to their related groups and *collaborations*. There are five basic function types Internal Logical Files(ILF), External Interface Files (EIF) in data functions and External Inputs (EI), External Output (EO) and External Query (EQ) in transaction functions. The elementary variable elements are known Data Element Type (DET) and elementary operations are termed as Process Element Type (PET). The complexity levels are calculated as user identifiable groups among DETs and are termed as Record Element Type (RET) in Transaction Function and the total count of ILF & EIF as File Type Referenced (FTR) in Transaction Function, similarly the user identifiable collaborations that exist in a process are used to determine the operational complexity termed as Logical Collaboration Segments (LCS). The total sum of the high, medium and low coefficients of all operations is the unadjusted function point count. There are 14 General System Characteristics used to ascertain the Adjusted Function point count, which is known as Value Adjustment Factor. This is used to calculate the complexity of environment, task and language and tweak the final count for the particular environment.

Deviations from IFPUG Standards

Deriving FP Count

Standard IFPUG

IDENTIFY FP RANGE					IFPUG FP REFERENCE			
EI - Value Ranges					EI - Progressive Index			
		DET					DET	
FTR		1-4	5-15	16+	FTR	1-4	5-15	16+
<2	1	Low	Low	Average	1	3	3	4
2	2	Low	Average	High	2	3	4	6
>2	3	Average	High	High	3	4	6	6

- Identify the Low, Average, High functions in EI, EO,EQ in transaction functions and ILF,EIF in data functions based on DET with respect to RET/FTRs.
- Count the sum of Low, Average and High counts in total for EI, EO, EQ, ILF, EIF function types.
- Multiply total sum for each function type by the Function Point Reference Index of Low, Medium and High FP Count.
- Count the sum total of all types to arrive at Unadjusted FP Count and multiply by VAF to obtain the Adjusted FP Count.

Progressive FP Count

IFPUG FP REFERENCE					Progressive FP Coefficients			
EI - Value Ranges					EI - Progressive Index			
		DET					DET	
FTR		1-4	5-15	16+	FTR	1-4	5-15	16+
<2	1	3	3	4	1	0.60	0.19	0.24
2	2	3	4	6	2	0.60	0.25	0.35
>2	3	4	6	6	3	0.80	0.38	0.35

- Identify the Low, Average, High functions in EI, EO,EQ in transaction functions and ILF,EIF in data functions.
- Multiply DETs by the Function Point Reference Coefficients based on Low, Average & High counts in EI, EO, EQ, ILF, EIF respectively.
- Count the Integrated Process Flows (IPF) in each function and multiply by the Coefficients of IPF table.
- Sum up the total value of functions by EI, EO, EQ, ILF, EIF, IPF and deduct from Reuse FP count to arrive at Unadjusted FP Count.
- Multiply by VAF to obtain the Adjusted FP Count.

Coefficient Index

IFPUG FP REFERENCE

	El - Value Ranges		
		DET	
FTR	1-4	5-15	16+
<2 1	3	3	4
2 2	3	4	6
>2 3	4	6	6
Range	5	16	17

Progressive FP Coefficients

	El - Progressive Index		
		DET	
FTR	1-4	5-15	16+
1, 2	0.7500	0.2000	0.2500
3, 4	0.7500	0.2667	0.3750
5, 6	1.0000	0.4000	0.3750
Range	5	16	17

We use coefficient index to obtain an accurate estimate and also to avoid the static range of FP values and provide a dynamically computed function point value based on the IFPUG FP Reference Index.

To illustrate the difference the FP values are plotted below

Type	DET	FTR	IFPUG UFP	Progressive UFP
EI	1	2	3.00	0.75
EI	2	2	3.00	1.50
EI	3	2	3.00	2.25
EI	4	2	3.00	3.00
EI	5	2	4.00	1.33
EI	6	2	4.00	1.60
EI	7	2	4.00	1.87
EI	8	2	4.00	2.13
EI	9	2	4.00	2.40
EI	10	2	4.00	2.67
EI	11	2	4.00	2.93
EI	12	2	4.00	3.20
EI	13	2	4.00	3.47
EI	14	2	4.00	3.73
EI	15	2	4.00	4.00
EI	16	2	6.00	6.00
EI	17	2	6.00	6.38
EI	18	2	6.00	6.75
EI	19	2	6.00	7.13
EI	20	2	6.00	7.50
EI	32	2	6.00	12.00

Given a scenario of a user input form with 4 DETs and 2 FTR in a function, the UFP count is 3FP as shown in the table above, 2 DETs will result in 1.5 FP in Progressive FP instead of 3 in standard FPA. The immediate benefit is that there is greater accuracy though it results in a lower count. The primary benefit of having coefficient index is there is no high value cut off. A user input page having 32 DETs and 2 FTRs may be counted with complexity level of High and allocated an FP value of 6 as per IFPUG standard. In progressive FP the value is computed as 32 x 0.3750 =12 FP. But using standard IFPUG FPA a function with 16 DETs with 2 FTRs are considered as High value FP and assigned a static value of 6. This causes sizing problems in projects where there are many complex forms with consistently higher number of inputs/outputs than present in the standard IFPUG range.

The Progressive function point formula is computed as follows:

$$PFP = VAF \times \left(L_{\substack{ind. \\ coeff.}} \times \sum_{i=1}^{n} L_i + M_{\substack{ind. \\ coeff.}} \times \sum_{j=1}^{m} M_j + H_{\substack{ind. \\ coeff.}} \times \sum_{k=1}^{q} H_k \right)$$

VAF	Refers to the computed Value Adjustment Factor.
L_i	This represents each Low Complexity Function ranging from 1 to n for a scenario.
M_j	This represents each Medium Complexity Function ranging from 1 to m for a scenario.
H_k	This represents each High Complexity Function ranging from 1 to q for a scenario.

$L_{\substack{ind. \\ coeff.}}$	Individual Coefficient computed for the given Low Complexity Scenario.
$M_{\substack{ind. \\ coeff.}}$	Individual Coefficient computed for the given Medium Complexity Scenario.
$H_{\substack{ind. \\ coeff.}}$	Individual Coefficient computed for the given High Complexity Scenario

Process Flow

Process flows are operations performed within each data and transaction function. For most CRUD (Create, Read, Update, and Delete) operations, there may be simple process flows and hence it does not cause any estimation drawbacks. In larger projects with many business processes, rules and alternative flows, the lack of counting the process steps in a function leads to under estimation of effort and inaccuracy in estimating the size of a project.

Type	DET	FTR	PET	LCS	IFPUG UFP	Progressive UFP	IPF
EI	2	1	10	3	3.00	1.50	2.63

In a user login process flow there are only two inputs the username and password, 2 DETs however there may be 10 or more process calls. In a standard IFPUG, the FP count is 3 FP for any DETs below 4, 2 DETs results in lesser count of 1.5 for Progressive FP count, but when adding the process count with the actual progressive FP count, the total value is 4.13 FP as illustrated. The difference is that both the input variables and process steps are considered in the equation thus gives a better sizing estimate and reveals the true effort involved.

It is mandatory to list the steps in the process flow which may be followed up with a simple activity diagram or flowchart and this provides a clear blueprint on the approach and ensures that the developer is provided with a guideline to follow.

The process steps are as follows:

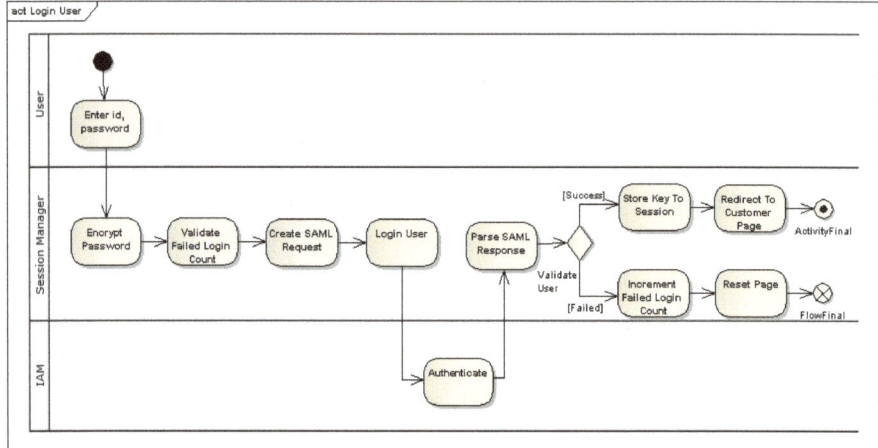

1. Encrypt Password	6. Parse SAML Response
2. Validate Failed Login Count	7. Store Key to Session
3. Login User	8. Redirect to Customer Page (or)
4. Create SAML Request	9. Increment Failed Login Count
5. Authenticate	10. Reset Page

We have 3 swim-lanes which are identified as LCS participants (User, Session Manager, IAM).

It is advisable to only count the computational steps which are represented by function calls in a sequence diagram and is traceable down to the code level.

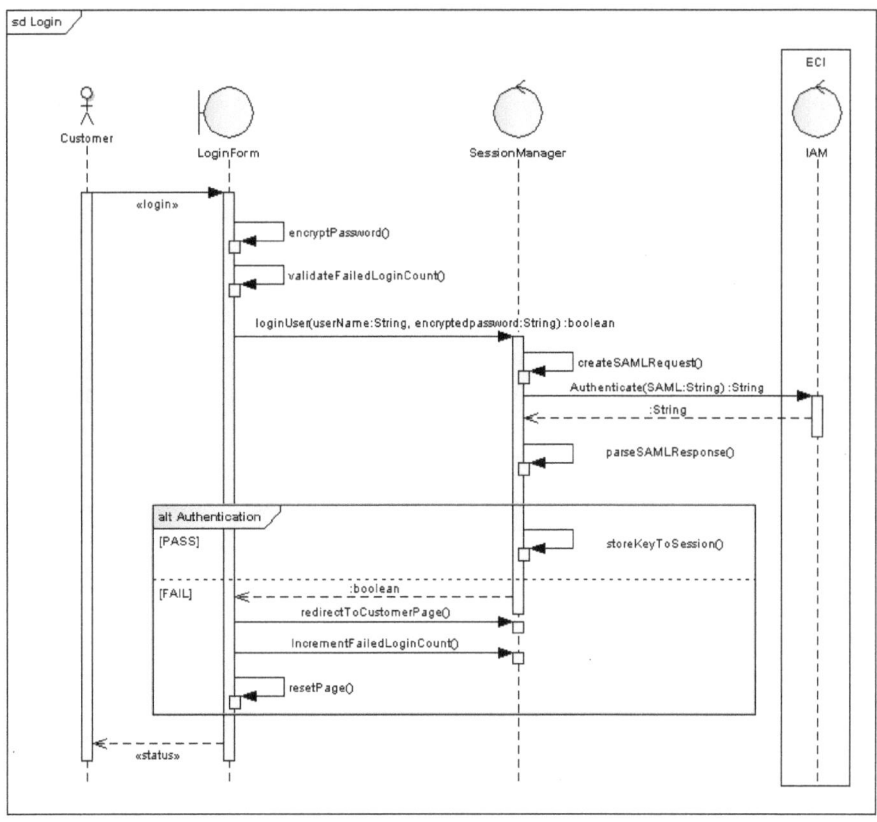

This is will be explored further in Automated Progressive Function Point Analysis. In an agile workforce there will be high level goals or parent user story often called an Epic, this will further be summarized by the child user story, which can be depicted using an Activity diagram, and coding mapped to the sequence diagram to provide end-to-end traceability. The blueprints are optional but advisable to use at least a flowchart for providing a better understanding to the developers concerned. Using a blueprint enhances a project by following a *Model Driven Development* approach. Using tools such as Enterprise architect enables the auto generation of most of the code freeing the developers to focus on tasks that are more important.

The Fundamental Concepts of Progressive FPA

Progressive Function Point measure uses coefficient index based on IFPUG standard function point reference values to arrive at an accurate function point count. Progressive FPA estimates the actual count of elementary elements and process steps in each function. Progressive function point reference is useful for accurate sizing, incorporating workflows, functional reuse and can relate to the actual effort involved since it takes into consideration all the variables in the function and the process steps involved. Progressive Function point estimation accounts for both data and transaction functions with integrated process flows.

Progressive FPA: Integrated Process Flow (IPF)

Integrated process flow comprises of the user recognizable, elementary non-repetitive process steps, which are present in both data and transaction functions and may include one or more process steps within respective logical groups inside the boundary of the application and may include references to External Component Interfaces to reuse functionality in other application(s) and to accomplish a desired outcome.

Progressive FPA: External Component Interfaces (ECI)

External Component Interfaces are integrated process flows of other applications. ECI are external functions referenced by applications to implement some desired functionality by reusing functions outside their application boundary.

Progressive FPA: Process Element Type (PET)

Process Element Type is user recognizable elementary computational process steps to manage, process or present data and control information within each data or transaction function. The primary intent is to accomplish a desired behavior through one or more activities within the application boundary, which is unique to the application and may include calls and external references to other External Component Interfaces (ECI) outside the application boundary.

Simple Definition

The list of operations or activities performed by a function or multiple group of functions is termed the process steps and used to deliver a desired result or business function as stated in the use case or user story. They are computationally relevant when they perform some action that delivers a unique outcome in terms of mathematical operation, logical operation, calculation or business function that may influence a change in its state. Ideally, each function will perform one unique action and follow the single responsibility principle, so there is better readability in the code, as it implements *Clean Code* practices. The parent process may further invoke multiple function calls to obtain the desired behavior or result where each function follows a single responsibility pattern. Count every process step performed within the boundary of the application and count only one-step for each external call or reference to External Component Interfaces even if all process steps are outside the boundary of the application. The External Component Interfaces are references to external application processes counted as PETs within their own application boundary hence may denote reuse. The process flows may be presented in UML through activity diagrams or sequence diagrams to adhere to a well-articulated architecture blueprint and thus both UML diagrams and wireframes may be used to add clarity to design and implementation. It is advisable to list the steps involved in each process flow for validation purposes.

Counting Rules
Count one process step for each computational operation performed by the function. Count one process step for each child function referenced by the parent function to accomplish a computational task.
Count one process step for any operation or group of computational operations processed outside the boundary of the application in a single external reference call to obtain the desired result.

Identifiers
Identify list of activities; try to identify each computational process steps within each function type. Each operation or activity should define a separate function to follow the single responsibility pattern. Decisions are not counted rather the function flow of all outcomes are counted. Identify both internal function calls and external references to any other components and separate the ECI from the process flow, which may include reusable components or utilities. Identify the functions performed inside the boundary of the application and count one-step for each delegated operation.

Progressive FPA: Logical Collaboration Segments (LCS)
Logical collaboration segments are unique user recognizable logical groups or participants that may exist both within the boundary of the application for each data or transaction function, and relate to ECI with external participants as separate swim lanes which are outside the boundary of the application. LCS are depicted as swim lanes in BPM and can also be visualized using activity diagrams in UML.

Simple Definition

A functional flow leads to many process steps and traverses through one or more logical groups, participants or segments within the boundary of the application. Additionally each External Component Interfaces is counted as one group.

Counting Rules

Count one logical collaboration segment for each user identified logical group, participant or segment.

Count one logical collaboration segment for each unique External Component Interface in the process flow.

Identifiers

Process flows are divided into many logical groups or swim lanes. Trace the function calls within the application boundary for unique operational sub tasks. Look for operational flow outside the boundary which may be identified by the use of libraries and shared utilities to identify ECI Components.

PROGRESSIVE FUNCTION POINT ANALYSIS WORKBOOK

 ## Progressive FPA Workbook

The progressive FPA workbook is an open source excel file available freely on sourceforge.com and can be downloaded from the following location.

 http://sourceforge.net/projects/functionpoints/

The open source project was introduced with an effort to establish the work and ideas implemented across many projects for improvements and standardization of progressive count and workflow estimation techniques. The workbook is divided into many parts where each worksheet provides a unique feature as described below.

- **Project Details**: Provides summary of both IFPUG and Progressive FP count and project details.
- **Use Cases**: Is the key reference indicator to all data and transaction functions contained within the use case, user story.
- **Reuse**: This sheet accounts for application components that are reused with refactoring assessment and implement Adaption Adjustment Factor.
- **Data Functions**: Accounts for all Added, Changed or Deleted ILFs & EIFs in the application.
- **Transaction Functions**: Accounts for all Added, Changed or Deleted EIs, EOs & EQs in the application.
- **FP Progressive Index**: IFPUG reference card and Progressive FP Coefficient matrix used for calculation of Project FP, Cost & Schedule.
- **FP Calculation**: Counts of summary presented in data and transaction functions with VAF, Cost and Schedule Index.
- **Reports**: Provides reports for a release based on keyed in data for easy export to various formats, and can save values without the need of the calculation sheet.
- **Editable Reports**: Provide a non-protected editable view of reports so the data can be filtered and analyzed by using excel macros and tools.

Project Details

FUNCTION POINT ANALYSIS

Version 3.0

Project Details	
Project ID	UP-01-1135
Project Name	University Portal Demo Project
Work Order	UP1135
Language	Apex/VisualForce
Client	ACME University
Start Date	10-Oct-14
End Date	10-Oct-15

Worksheet Summary	Progressive FPA	IFPUG FPA
FP Count	27	34
Reuse FPs	9	6
Process FPs	7	
Estimated Man Hours	66	
Total Cost	$11,244.19	

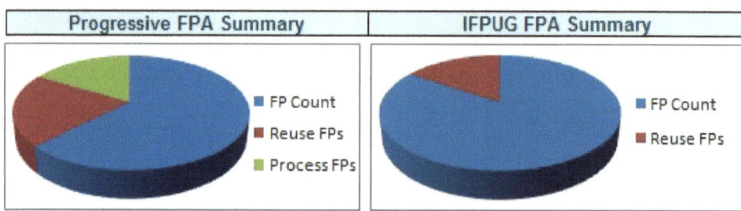

Progressive FPA Summary	IFPUG FPA Summary
FP Count / Reuse FPs / Process FPs	FP Count / Reuse FPs

Grant Of License	
License Plan	ENTERPRISE
License Number	2110140001
Project	
Company	

The summary sheet provides a quick overview of the project statistics, both IFPUG FPA count and Progressive FPA count from the calculation worksheet.

Here we enter the internal project related details such as ID, Name, associated Work Order, Language(s) used, Client, Tentative Start date and End date of this project. The second part highlights the project summary, which is taken from the FP calculation sheet, and is automatically computed as the data and transaction functions are being entered. The details shown here are the calculated FPs in the project, which comprises of the total FP count, the Reuse FP count, and the Process FP count, and the estimated man-hours for the project. Process FP is used to include Integrated Process Flows. As the estimation sheet provided uses high accuracy by implementing multiplicity factor instead of static ranges, it is possible to compute both estimated man-hours and total cost of the project.

The next section is the Licensing, which is issued to all who support the project, licensing helps to support the project and it is requested that all commercial projects may purchase a licensed copy of the sheet, which entities partner support.

Use Cases

PROGRESSIVE FUNCTION POINT ANALYSIS

USE CASES/USER STORIES/SCENARIOS

Reference ID	Description	UFP	Reuse FP
UC1001	Student Enrollment	5.28	0.72
UC1002	Upgrade static University website to WCMS	6.05	0.11
UC1003	Remove General feedback Form on University Website	1.75	0.00
UC1004	Refactor existing web components	0.00	6.43

There are four columns which consists of Reference ID, Description, Unadjusted FP Count (UFP) and Reuse FP. The user only enters the Reference ID and Description, which can be the use case or user story, and it is possible to reference each data and transaction function in the use case by the unique reference ID. You are required to provide a unique ID and description of the use case. Once you provide all the data and transaction functions in the use case/user story the Unadjusted FP count and Reuse FP count will be visible here to help you understand the workload in each UC/US. This is auto computed in the sheet as you enter the details of the work involved. This view helps you to assess the workload and assign and prioritize the tasks to your team with better insight.

Reuse

REUSEABLE APPLICATIONS/COMPONENTS

			Modification Required					
Reference ID	Description	FP Count	Design	Code	Testing & Integration	Reuse FP	Refactor Man Hours	Project Reference
UC1004	PDF Writer Component - Reuse Reporter	5	25%	50%	100%	2.75	8.25	TMW Platform - Rel AA - Reporter Component
UC1004	Emailer Component - Reuse NotifyMe	7	30%	60%	75%	3.68	11.03	Concorde Project, NotifyMe Component

Adaption Adjustement Factor (AAF=.4Design Modification×.5Code Modification x .5Testing & Integration)

The Reuse sheet applies the Adaption Adjustment factor to validate the effort required in integrating components or applications as reusable parts of a project. There are 9 columns in this sheet, the first column should be the unique Reference ID, and should not be the same as that of any other Use Case/User story. The second column is for description, this is for naming the component or application that will be reused. The third column is the FP count of the application or component being integrated into the project. The next three columns indicate the level of work required to transform the component or application to be integrated into the current project in terms of Design changes, coding effort, testing and integration effort. The Reuse FP is computed using the Adaption Adjustment Factor using the formula

AAF = 0.4 x Design Effort + 0.3 x Coding Effort + 0.3 x Testing & Integration

AAF was presented in an IFPUG conference, hence this effort may also calculated as Reuse FP with the standard IFPUG count.

Data Functions

This page consists of 3 parts Added, Changed and Deleted. These are functions counted in the persistence layer.

Added: Relate to creation of table or records and the associated process steps in the creation of any triggers, functions, procedures in the database.

ADDED

Logical Files	Reference ID	ILF/EIF	DET	RET	PET	LCS	Man Hours	UFP	UPFP	IPF	Reuse (Y/N)	Reuse %	Reuse FP
						New Unadjusted FP, Progressive FP, IPF Count							
Create Student Records	UC1001	ILF	10	2	3	2	16	7	3.68	1.80	Yes	10	0.56
Create Content Records	UC1002	ILF	15	2	3	2	22	7	5.53	1.80	No		
							38	14.00	9.21	3.60			0.55

DET Details	RET Details	PET Details	LCS Details	Reuse Details
1. Student ID. 2. Name. 3. DOB. 4. Address. 5. Parent Name. 6. Contact. 7. Course. 8. Enrollment Date. 9. Duration. 10. Subject	1. Student Records. 2. Course Record	1. Create Student Registration Trigger. 2. Validate Consistency and Completeness. 3. Save Records	1. Student. 2. System	New Tables, Triggers, Validations
1. Content ID. 2 Type. 3.Title. 4.Content. 5.Version. 6.Publication Date. 7.Submitted Date. 8.Author 9. Page Id. 10. Page Name. 11. Icon. 12. Title. 13. Layout. 14. Template. 15. Theme	1. Content Records. 2. Page Records	1. Create Content Validation Trigger. 2. Validate Consistency and Completeness. 3. Save Records	1.Editor. 2 Writer	New Tables, Triggers, Validations

The first column is used to describe the task, the second column is a dropdown and allows you to select the related Use Case/User Story. The third column is another drop down to indicate if its local store (ILF) or remote storage (EIF). The fourth and fifth column is the DET element count with RET, and the sixth and seventh column is PET and LCS. Once these details are provided the function point value is calculated automatically. The eight column is used to predict the man hours, the UFP count is the standard Unadjusted FP Count in column nine, and UPFP count is the Unadjusted Progressive FP count in column ten with IPF process count in column eleven. The reuse column twelve indicates if there is any function reuse present in the given function, where a part or whole of the operation and process may be reused, the next column thirteen shows the percentage of reuse and the Reuse FP Count is calculated in column fourteen. The next columns are used to show the actual DET, RET, PET, and LCS in the function so they may be verified and helps in the justification of the FP Count if there is a change from the projected and actual, this greatly helps to mediate concerns.

Changed: These are existing functions with a new change implemented, which may require the addition or modification of elements.

CHANGED

Logical Files	Reference ID	ILF/EIF	DET	RET	PET	LCS	Man Hours	UFP	UPFP	IPF	ILF/EIF	DET	RET	PET	LCS	Man Hours	UFP	UPFP	IPF
							Unadjusted FP, Progressive FP, IPF Pre Count									Unadjusted FP, Progressive FP, IPF Post Count				

DET Impact	RET Impact	PET Impact	LCS Impact	Reuse (Y/N)	Reuse %	Reuse FP	DET Details	RET Details	PET Details	LCS Details	Reuse Details

The first column is the description, and the second column is the Reference ID. The next four columns are the actual DET, RET, PET and LCS in the existing function and the FP count is calculated in the next four columns to denote the previous count, and the next four columns are the current count of DET, RET, PET & LCS in the function. The next four columns show the impacted DETs in the function, which may also be existing elements. The Reuse FP indicator is the next column and a percentage of reuse is provided in the column after to represent the total reuse FP in the changed FP count. The last five columns is used to justify the changed FP counts and Reuse FP Counts.

Deleted: The deleted function represents the element count before deletion of the function or operation concerned.

DELETED

Logical Files	Reference ID	ILF/EIF	DET	RET	PET	LCS	Man Hours	UFP	UPFP	IPF
		Unadjusted FP Count + IPF Count Post Count Before Deletion								
Delete Feedback Records	UC1003	ILF	5	1			6	7	1.84	

DET Details	RET Details	PET Details	LCS Details
1. Feedback_ID, 2. Guest_Name, 3. Email, 4. Comments, 5. IP Address	Feedback Records		

The description of the deleted function is provided in the first column. The Reference ID is provided in the next column. The next column identifies the data storage as local ILF or remote EIF type. The following four columns is used to denote the actual count of DET, RET, PET and LCS before deletion of the function to identify the deleted FP count. The last four columns list the elements for any validation or justification.

Summary: The summary sheet contain the subtotals of data function in the calculation sheet provided. The progressive FPA summary is the summary of the estimates only for progressive FP calculation and the IFPUG UFP summary calculate only the IFPUG standard count for the given function in the sheet. Both of the summary totals is used by the FP Calculation worksheet. The above Data Functions Progressive summary sheet show a summary of the ILF Count, EIF Count and

Progressive FPA Summary

	Added	Changed	Deleted	Reuse
ILF Count	12.26	0.00	1.84	0.55
EIF Count	0.00	0.00	0.00	0.00
Man Hours	38.43	0.00	5.53	

Total Added FP	12.26	0.55
Total Changed FP	0.00	0.00
Total Deleted FP	1.84	
Total Process FP	3.60	
Total Reuse FP	0.55	
Man Hours	43.96	

IFPUG UFP Summary

Total Added FP	14.00
Total Changed FP	0.00
Total Deleted FP	7.00

Man-hours required along with the Added, Changed, Deleted, Reuse and Process FP counts. The IFPUG Summary sheet calculates the sum total of all Added, Changed and Deleted FP Counts.

Transaction Functions

Added: The addition of new UI's or input/output operational functions generally fall under the new added transaction category.

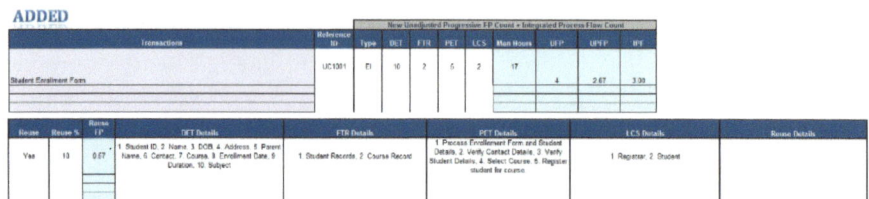

The first row describes the function, the second column is a dropdown selection of Reference Id, and the following third column is a selection box to select the type of operation (EI, EO, EQ), the next four columns are used to input the number of DET, FTR, PET, LCS in the function. The next four columns after display the computed values of estimated man hours, Unadjusted FP based on IFPUG model, Progressive Unadjusted FP Count, and Integrated Process Flow count. The next column is an option to show reuse if any part has been reused, with the total reuse percentage to calculate the Reuse FP count. The last four columns list the elements counted for validation and justification process and the final column describes the reuse in detail.

Changed: These are existing transaction functions with some changes to be implemented, which may require the addition or modification of elements or process flow.

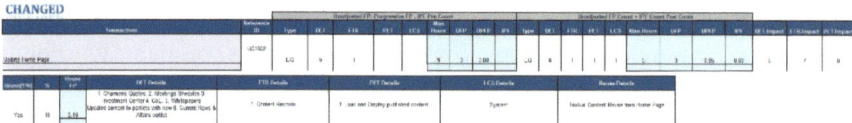

The first column describes the transaction which may be a UI or input/output operation, the second column is the selection dropdown and is the reference identifier to the use case or user story. The third column is a selection which identifies the type of function(EI, EO, EQ), the next four columns provide the existing element count for DET, FTR, PET & LCS, and computes the man-hours, Unadjusted FP Count, Unadjusted Progressive FP Count and Integrated Process Flow count. The next four columns provide the modified count of DET, FTR, PET & LCS, this may include the newly added elements may have removed certain elements. The next four columns depict the computed total of man hours, Unadjusted FP count, Progressive FP Count and IPF Count followed by the impacted element columns, which describe impact on the number of DET, FTR, PET & LCS to show the actual impact. If 8 DETs were added & 8 were removed, it will be difficult to address the change, since the DET Count remains the same. The impacted elements enable a clear view of what were the changes and the same justification of elements is provided in the last four of five columns. Reuse is the last column which also helps to add the percentage of reuse based on the impact for justification and validation purpose.

Deleted: The UI or input/output operations count that is taken prior to the removal of the function concerned is the Deleted FP Count.

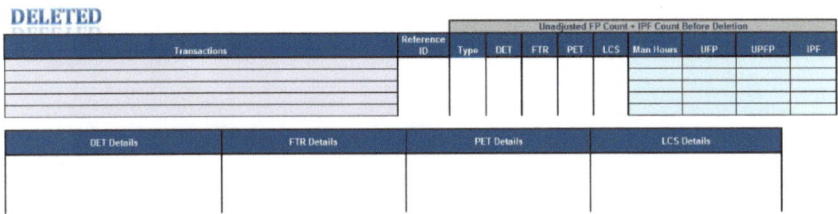

The first column is used to describe the deleted function, the next column is the use case reference ID. The third column is used to denote the type of function (EI, EO, EQ). The next four columns are the counts of elements DET, FTR, PET, & LCS, for which the man hours is computed in the adjacent column with the Unadjusted FP count, Unadjusted Progressive FP Count, and Integrated Process Flow count. The remaining four columns are used to list the actual DET, FTR, PET & LCS.

Summary: The summary sheet is the subtotals of data function in the calculation sheet provided

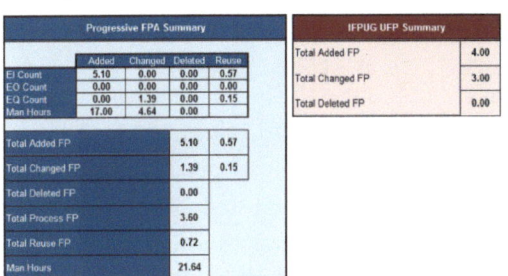

The FPA summary is the sub-total of the operations in the worksheet. There are two summaries one for progressive FP calculation and the other for Standard IFPUG count for the given functions in the sheet. Both of the summaries are used by the FP Calculation worksheet. The above Data Functions Progressive summary sheet show a summary of the EI, EO and EQ Count, Man-hours required along with the Added, Changed, Deleted, Reuse and Process FP counts. The IFPUG Summary sheets calculate the sum total of all Added, Changed and Deleted FP Counts.

FP Progressive Index

IFPUG FP REFERENCE

EI - Value Ranges

FTR		DET 1-4	5-15	16+
<2	1	3	3	4
2	2	3	4	6
>2	3	4	6	6
Range		5	16	17

Progressive FP Coefficients

EI - Progressive Index

FTR	DET 1-4	5-15	16+
1, 2	0.7500	0.2000	0.2500
3, 4	0.7500	0.2667	0.3750
5, 6	1.0000	0.4000	0.3750
Range	5	16	17

EQ - Value Ranges

FTR		DET 1-5	6-19	20+
<2	1	3	3	4
2,3	3	3	4	6
>3	4	4	6	6
Range		6	20	21

EQ - Progressive Index

FTR	DET 1-5	6-19	20+
1, 2	0.6000	0.1579	0.2000
3, 4	0.6000	0.2105	0.3000
5, 6	0.8000	0.3158	0.3000
Range	6	20	21

EO - Value Ranges

FTR		DET 1-5	6-19	20+
<2	1	4	4	5
2, 3	3	4	5	7
>3	4	5	7	7
Range		6	20	21

EO - Progressive Index

FTR	DET 1-5	6-19	20+
1, 2	0.8000	0.2105	0.2500
3, 4	0.8000	0.2632	0.3500
5, 6	1.0000	0.3684	0.3500
Range	6	20	21

ILF - Value Ranges

RET		DET 1-19	20-50	51+
<2	1	7	7	10
2,5	5	7	10	15
>5	6	10	15	15
Range		20	51	52

ILF - Progressive Index

RET	DET 1-19	20-50	50+
1,2	0.3684	0.1400	0.1961
3,5	0.3684	0.2000	0.2941
6,8	0.5263	0.3000	0.2941
Range	20	51	52

EIF - Value Ranges

RET		DET 1-19	20-50	51+
<2	1	5	5	7
2,5	5	5	7	10
>5	6	7	10	10
Range		20	51	52

EIF - Progressive Index

RET	DET 1-19	20-50	50+
1,2	0.2632	0.1000	0.1373
3,5	0.2632	0.1400	0.1961
6,8	0.3684	0.2000	0.1961
Range	19	50	51

IPF - Value Ranges

LCS		PET 1-5	6-19	20+
<3	2	3	3	5
2,5	5	3	5	7
>5	6	5	7	7
Range		6	20	21

IPF - Progressive Index

LCS	DET 1-5	6-19	20+
1,2	0.6000	0.1579	0.2500
3,5	0.6000	0.2632	0.3500
6,8	1.0000	0.3684	0.3500
Range	6	20	21

Currency	Dollar	Workbook Customization	World Currency appiled to worksheet
Hours/FP	3.0	Effort Estimation	Computed Hours per FP
Cost/FP	$350.00	Cost Estimation	Cost per new FP computed
Cost/Reuse FP	$200.00	Reuse Cost	Charges for Reused/Integrated components

The FP Progressive Index is based on IFPUG standard values and their coefficients. Workbook related configurations such as currency, cost per FP count, and cost of Reuse FP for reuse and integration costs can be adjusted. The value ranges and coefficients are not to be touched. To tweak these values as described in chapter 4 you may edit the columns in blue in IFPUG reference tables. The dark blue columns (Column 2, Rows 4-6 and Colum 3-5, Row 7) are used in the actual formula to define the ranges, great care should be taken in editing these values. To train the sheet the FP counts may be changed in the light blue columns (Row 3-5, Columns 4-6). These adjustments are applied only for training the worksheet based on existing projects for CPMG audits and performance tweaking over time, do not change these values as they will deviate from IFPUG standard.

FP Calculation

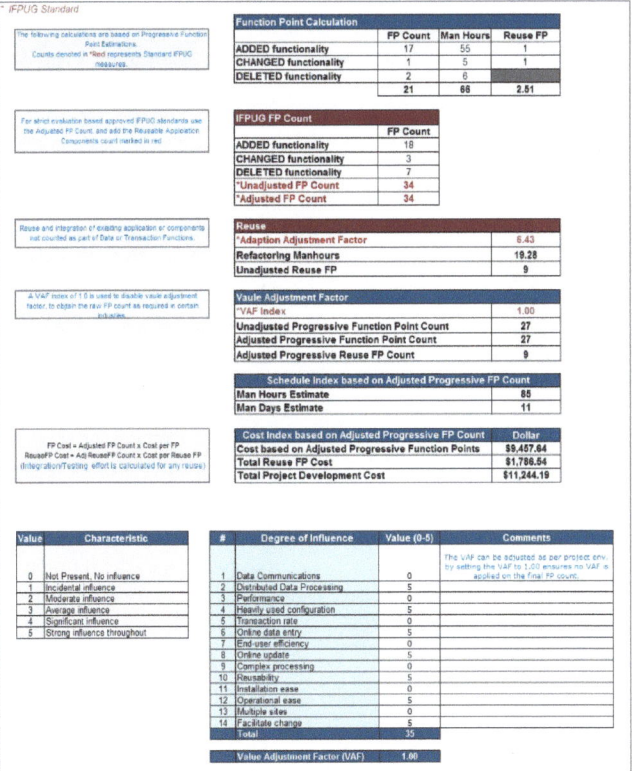

The first section is subtotal for all added, changed and deleted functionality taken from the data and transaction summary sheets. This provides an unadjusted progressive FP count and man-hours required for the project with Reuse FP Count. The second section if the Standard IFPUG calculations, which displays the data and transaction added, changed and deleted function counts with unadjusted FP count and adjusted FP count based on the VAF parameters calculated below. The third section measures the reuse total based on AAF, and the refactoring man-hours, the Unadjusted Reuse FP count includes the AAF and functional reuse count. The fifth section shows the computed VAF. A VAF total of one indicates that the VAF is not applied and it is the "Raw" FP count, technically called the unadjusted FP count. The next three rows show the progressive unadjusted FP count, adjusted FP count and Reuse FP count. The sixth section is used to compute the schedule index and calculates the estimated man hours and man-days for the project. The seventh section is the cost index based on the progressive FP count and displays the cost of total FP counts, Reuse FP counts and total project development cost. The section below that is used to compute the VAF as explained in the 1st chapter based on the 14 General characteristics.

Reports

Reports are useful to maintain the data outside the datasheet and allow the export of data. This sheet provides the added, changed and deleted functions in data and transaction functions. It is possible to store the same as text, csv, pdf or any other format required for easy exchange.

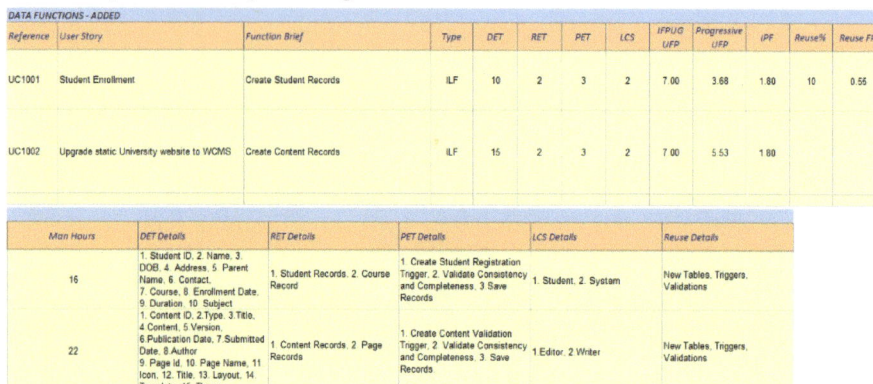

Export to PDF

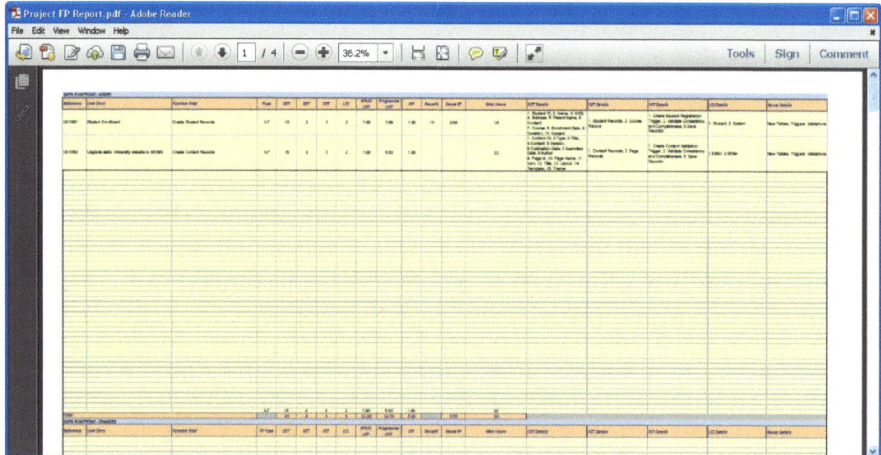

1. Ctrl + P or Print from the Office menu.
2. Select Printer Name: Microsoft Document Image Writer
3. Click Properties
4. Select Page Tab
5. Page Size: Custom (Width: 28.7, Height: 35, Landscape)
6. Save As PDF (Download free Microsoft Office PDF Converter Plugin)

Editable Reports

The editable reports section was added for easy presentation and analysis of data. It can be used to order, sort or display data in any format desirable.

Order By Reference ID

This lists all the data and transaction functions in ascending or descending order.

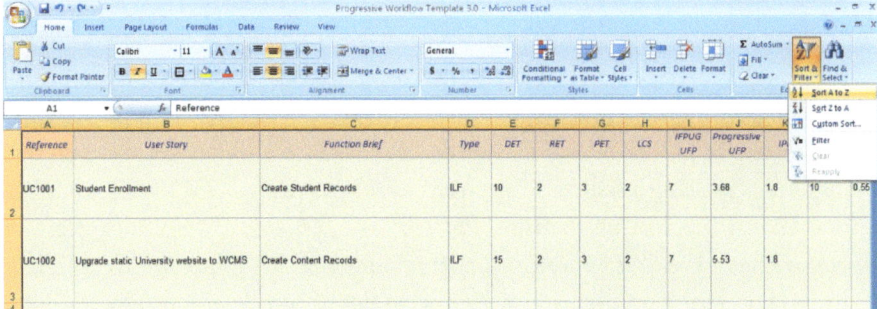

1. Select all the columns shaded in yellow along with the header.
2. In Homepage tab, select Sort & Filter
3. Sort A to Z

Filter and Sort

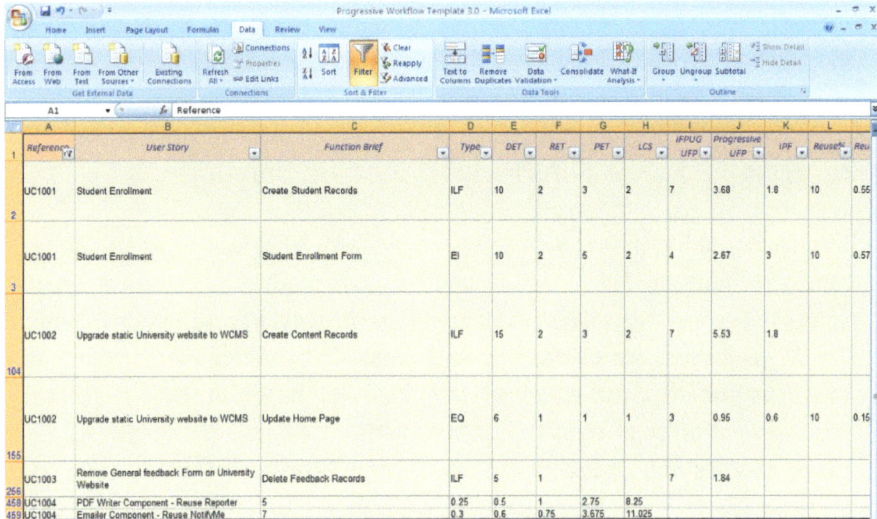

1. Select all columns shaded in yellow along with header.
2. Click Data Tab
3. Click Filter
4. Click on first column filter dropdown and deselect blank
5. Click on first column filter dropdown and click Sort A to Z

PERFORMANCE AUDITING & TWEAKING

CMMI Assessment

CMMI (Capability Maturity Model Integration) is a Process Model Framework for process-improvement developed by Software Engineering Institute (SEI), Carnegie Mellon University (CMU). It is a structured and systematic collection of best practices for process-improvement.

The SEI Assessor will require the stepwise account of the calculation process, the method of arriving at the FP counts and the approach taken by the team in arriving at the Project FP Count. This will require training of the sheet from previous projects. The FP ranges of High, Medium and Low may need to be adjusted based on the actual data from previous projects. This may be counter to the IFPUG philosophy of having one FP count across all domains and projects. To adjust the values there are two approaches. The first approach is bottom up approach where we try to build a new truth table based on user analogy on what constitutes low, medium and high effort based operations, which the worksheets are trained till the optimum values are met for the platform. The second approach requires the historical data on projects from a number of releases where the sheet undergoes several tests with respect to range and distribution. The low, medium and high effort operations are tested against the values in the sheet and are trained over a number of projects to identify the optimum values. Both require considerable effort and commitment from teams on a continued basis. Only trained sheets with proven values are permissible for projects.

Altering FP Range For Practice Specific Domains

By analyzing the application domain, and platform it is possible to arrive at estimation range for defining High, Medium and Low complexities and the corresponding FP values. These are the values provided for each complexity scenario for computing FP. The FP values for each platform or domain has to be determined by the organization and must be measurable work estimates and historical data from the project management office or concerned department. The data derived extends beyond the range of VAFs and calculates the unadjusted FP values for each given complexity scenario.

To arrive at these estimates we have to identify how much time and effort is required to model and implement an application in a platform or domain. There are two possible approaches; one is the Build-up approach where we arrive at a value based on analogy and experience in the domain, where previous data on that platform or domain may be sparse or non-existent. The other approach is a Revise approach, where we have some established means of computing the cost and schedule index, and have historical data as reference of completed projects and can be used to train the worksheet and compute the FP ranges for a given platform.

The Build-Up Approach

The build-up approach requires the construction of a truth table based on similar projects, which will help in determining the total work effort and value ranges required for the given domain. The truth table will need to satisfy the following conditions:

For Data functions we identify the database requirements, how much engagement is required here. Apart from modeling the database, the additional requirements for developing triggers, functions and procedures are addressed including maintenance and management requirements for ILF. For EIF we need to consider the technology used for storing the data externally using the given interfaces, middleware, and the other factors that govern the data integrity, availability, concurrency, reliability, and scalability needs.

Create a truth table as follows

- Identify tasks that can be completed in a day or two, the second set would have work that can be completed in a week and third set for work that would take more than a week.
- Shortlist all work that can be completed in the given timeframe set in three separate sheets for data and transaction functions.
- Identify the new ranges for Low, Medium, High complexity scenarios from the three sheets.
- Determine the average Count for Low, Medium and High scenarios as the new FP Count.
- The new FP count will be identified for all data and transaction functions after which GCD is applied to all the FP Counts for normalization.
- Since the ranges were identified using given time frame, the next step is to find the average time for delivering the given FP count.
- Finally derive the cost per FP.

Validate the sheet with number of projects and tweak the values on incremental basis over many project to arrive at the final domain specific value set.

The Revise Approach

The revise approach is more suitable where there is availability of historical data on the functions being performed in a particular platform to ascertain the costs or availability of cost matrix chart to compute the costs. In most cases, release wise info on both data and cost matrix charts should be available from the respective project departments especially where there is already an established FPA practice.

To perform this evaluation we will need to obtain the complexity ranges and unadjusted FP values provided for all data and transaction functions. The next step involves analyzing data from the previous releases. The data once obtained can be used to build a distribution chart for data and transaction functions to illustrate if the project ranges fall under the IFPUG estimation ranges or exceeds the computational bounds. To assess the counting practice effectively it is required to collect sample data for the past few years or as many releases as possible.

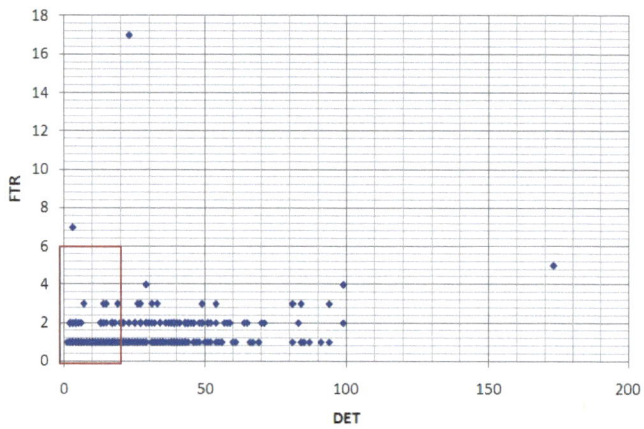

The above chart was drafted by analyzing the data of Release P -Release Z of the TMW platform with two years data. It was evident from the input, that the value range for given FP count were too less and unevenly distributed outside the scope, where all FP Count were simply measured as high value data functions, which is accurate as per IFPUG model, but given the DET, RET count the dispersion graph was able to clearly show the disparity that existed. Since the project cost was associated with FP count, this project was incurring great losses to the company.

By creation of probability graph it was more evident that only 15% of the actual data falls within the standard ranges as shown below.

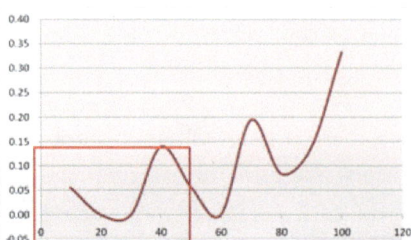

The problem was mitigated using progressive FP count, however the worksheet should be trained to reflect the actual domain values. The following would be the steps to tweak the worksheet ranges and FP counts.

- Collect sample data of existing projects for as many years as possible.
- Analyze data and transaction function separately.
- Analyze each type separately.
- Define the existing boundary (IFPUG reference ranges).
- Check for consistency in ranges.
- Identify low, medium and high complexity operations and their associated ranges.
- Tweak ranges to the most probable value ranges.
- Identify the average time frame for low, medium and high complexity for each type.
- Identify the Integrated process flows by identifying the activities involved for each operation in the architecture document HLDs in the project or through auto generated sequence diagrams to relate to each step involved. Use tools such as eclipse or reverse engineering the project to UML models.
- Compute the new FP count by calculating the GCD of computed time frame values.
- Compute the man hours required per FP.
- Compute the new cost per FP.
- The new sheet is to be tested over next several releases.

The final values for the domain is published to the global repository on progressive FP site, which further allows all other companies to adapt to the new standard or improve on the results based on their findings.

Glossary

AFP	Adjusted Function Point
APFPA	Automated Progressive Function Point Analysis
DET	Data Element Type
ECI	External Component Interface
EI	External Input
EIF	External Interface File
EO	External Output
EQ	External Query
FPA	Function Point Analysis
FTR	File Type Referenced
GSC	General System Characteristics
IFPUG	International Function Point Users Group
ILF	Internal Logical File
IPF	Integrated Process Flow
LCS	Logical Collaboration Segments
PET	Process Element Type
PFPA	Progressive Function Point Analysis
RET	Record Element Type
UFP	Unadjusted Function Point
UI	User interface
VAF	Value Adjustment Factor